Design for Mechanical Fatigue

A monograph covering:
- Material Fatigue Properties
- Fatigue Failure Prediction
- Complex Stress Loading
- Partial Fatigue Damage

Carl F. Zorowski
Mechanical Design Engineering
Monograph IV

Design for Mechanical Fatigue

Copyright 2016
All Rights Reserved

Design for Mechanical Fatigue

Table of Contents

Chapter 1 – Material Fatigue Properties
- Fatigue Failure Process — 1
- Fatigue Failure Examples1 — 2
- Fatigue Strength Testing — 4
- S-N Test Result Diagram — 5
- Fatigue Strength Equation — 6
- Approximate Endurance Stress — 7
- Endurance Modifying Factors — 8
- Surface Factor — 9
- Size Factor — 9
- Reliability Factor — 10
- Temperature Factor — 11
- Stress Concentration Factor — 11
- Miscellaneous Effects — 12
- Sample Application — 13
- Problem Solution — 13

Chapter 2 – Fatigue Failure Prediction
- Fluctuating Stress State — 17
- Stress Time Behavior — 18
- Mean/Fluctuating Stress Diagram — 19
- Fatigue Failure Theories — 20
- Modified Design Theory — 21
- Sample Problem — 22
- Check Modified Goodman — 28
- Fluctuating Torsional Fatigue — 30
- Torsion Fatigue Design Criteria — 31

- Torsional Fatigue Formula   32

Chapter 3 – Multiple Stress Loading
- Combined Loading Analysis   33
- Generic Rotating Shaft   35
- Element Stresses at A and B   36
- Element Stresses at C and D   37
- Mean and Alternating Stresses 38
- Graphic Representation   39
- Sample Problem   40
- Mean/Alternating Stress Values 42
- Von Misses Stresses   43
- Goodman Diagram   43

Chapter 4 – Partial Fatigue Damage
- Damage Accumulation   47
- Palmgren-Miner Theory   48
- Sample Problem   49
- Log-Log SN Diagram   49
- Part 1 Solution   50
- Part 2 Solution   52
- Part 3 Solution   52
- Mason Modification   55
- Part 1 Solution Modified   56
- Part 2 Solution Modified   57
- Part 3 Solution Modified   58
- Summary Results   58

Design for Mechanical Fatigue

Preface

Design for Mechanical Fatigue deals with the complex process of failure of metals when subjected to variable repetitive loading over large numbers of operating cycles. Failure under these conditions can occur at loading levels significantly below the static strength capability of the material and can be catastrophic. The use of the word "fatigue" is appropriate in that metal behaves as if it gets "tired" of being loaded repetitively and simply undergoes a brittle fracture even though it was originally very ductile.

The progression of a fatigue failure is a complex series of additive events. This complicates its description by simple mathematical models that permit its prediction for design purposes. The information presented here is representative of the models and their application that are generally accepted as providing a reasonable estimate of predicting fatigue behavior.

Chapter 1 covers the fatigue properties of materials. This includes testing for fatigue "strength", how life cycles are dependent on loading, modeling this relationship and consideration of physical factors that reduce ideal fatigue "strength" test results.

Chapter 2 deals with fatigue failure prediction under repetitive one-dimensional variable normal stress or shear stress states. This covers how these repetitive loadings are classified and establishing a failure theory based on the load classification, material properties, physical observations and conservative limitations. A detailed solution to an example problem is presented predicting factors of safety and total cycle life for a specific set of operating conditions.

Chapter 3 deals with how the one-dimensional stress state model is extended to cover multiple simultaneous stress loadings. This is accomplished with a six-step process that makes use of Principal and Von Misses stresses together with the one-dimensional failure theory. A detailed description of the application of this process is also included.

Chapter 4 considers the issue of estimating partial fatigue damage and damage accumulation when loading is changed during life cycle operation. The Palmgren-Miner Summation Theory and the Mason Modification of this approach is presented together with an example of their application to illustrate how this issue can be addressed and how the results vary.

As indicated in previous monographs the goal of Design for Mechanical Fatigue is to provide a basic understanding of the process of fatigue and the models and design criteria that are appropriate in the engineering application of estimating life cycle behavior. It is not intended to be a textbook or comprehensive reference source. Its purpose is to assist the previously acquainted reader in recalling relevant content and to provide a concise summary of the subject's accepted knowledge base for those acquiring the information for the first time.

About one third of the monograph's content consists of graphical and mathematical presentations. These have been included to help explain and provide for a more complete understanding of the subject content. The time spent in integrating these visual presentations with the written text will deepen the reader's depth of comprehension of the subject.

The material contained in this monograph is abstracted from a course supplement (*Design for Strength and Endurance,* ISBN 0-973126-1-3) available in PDF format at www.designforstrength.com. This supplement was prepared for a Mechanical Design Engineering course offered at North Carolina State University by the author.

Carl F. Zorowski
Cary, NC
November, 2016

Design for Mechanical Fatigue

Chapter 1 – Material Fatigue Properties

Chapter 1 of Design for Mechanical Fatigue deals with the properties of materials subjected to cyclic fatigue loading. Included are the characteristics of fatigue fracture, effect of stress level on cycles to failure, the endurance limit, infinite life and physical factors that reduce fatigue strength.

Fatigue Failure Process

The process of fatigue failure due to excessive repetitive stress cycling is quite different from static yielding or ultimate fracture. Fatigue is initiated at some material or surface imperfection like a grinding scratch or sudden change in geometry. A key way or small hole can give rise to a stress concentration and the creation of a small crack. This increases the stress concentration that further induces continuing growth of the crack.

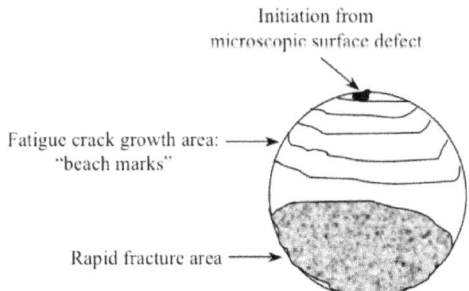

Figure 1-1 Process of Fatigue Failure

As the crack grows the stressed area decreases and nominal stress levels increase. This is usually accompanied physically by what are called "beach marks" visible over some portion of the final fractured surface, see Figure 1-1.

Total failure and fracture separation occurs suddenly with virtually no warning when the remaining stressed area can no longer withstand the external loading. The final fracture surface area appears like that of a brittle material in tension even though the design material was actually ductile. The phenomenon of fatigue failure occurs at stress levels below the ultimate static strength of the material.

Fatigue Failure Examples

The upper left of Figure 1-2 shows the fracture surface of a shaft subjected to cyclic torsion. Zones one and two illustrate the progression of the continued growth of the crack from the point of initiation. When the loaded stress area has been reduced to about two thirds of the original shaft cross section the torsion loading produces catastrophic separation. The middle upper photograph shows a fatigue failure of a crankshaft that clearly exhibits beach marks emanating from a small lubrication hole. The top right is a bicycle sprocket that failed due the stress concentration on the right that lead to the crack region working across the cross section to the final brittle fracture appearance on the left side. The bottom left is a typical fatigue fracture surface in a spring that is the result of reversed torsion as the spring is compressed and relaxed.

Design for Mechanical Fatigue

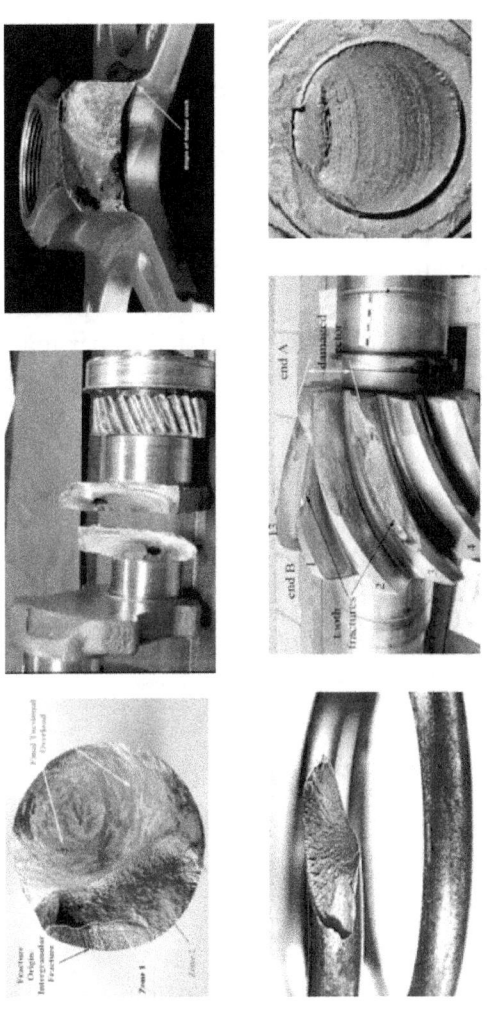

Figure 1-2 Fatigue Fracture Examples

Design for Mechanical Fatigue

The middle bottom illustrates how fatigue failure can occur on gear teeth subjected to repetitive use cycling that might originate from machined tooth surface imperfections. The final photo illustrate how a change in geometry from the reversed key at the top can be the source to create the well defined beach mark crack propagation prior to final catastrophic fracture at the bottom.

Fatigue Strength Testing

The fatigue strength properties of materials are determined by subjecting standardized test specimens to either cyclic cantilever or pure bending as depicted in Figure 1-3. These test specimens are rotated under load at high speed and the revolutions to failure are

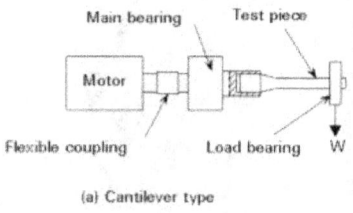

(a) Cantilever type

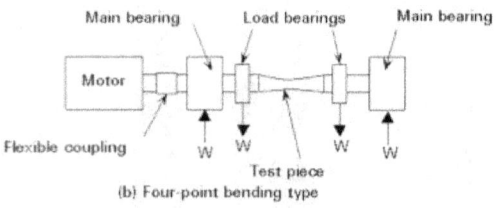

(b) Four-point bending type

Figure 1-3 Fatigue Test Arrangements

measured under different values of repetitive reversed normal stress produced by the static bending loads. Special care must be exercised in the machining and preparation of test specimens to minimizes any surface imperfections. Numbers of tests are required at a given load due to the statistical nature of the failure process. These are referred to as S-N tests.

S-N Test Result Diagram

Illustrated in Figure 1-4 is a typical S-N diagram in which fatigue strength or the stress that causes fatigue fracture is plotted as a function of the cycles of reversed loading on a log-log scale. The plotted points are indicative of normal statistical scatter. The diagram consists of several characteristic regions. One is characterized as low cycle, from zero to a thousand, and a second high cycle region from a thousand to one million cycles or more. The other characterization is the cycle range for finite life contrasted to that of infinite life.

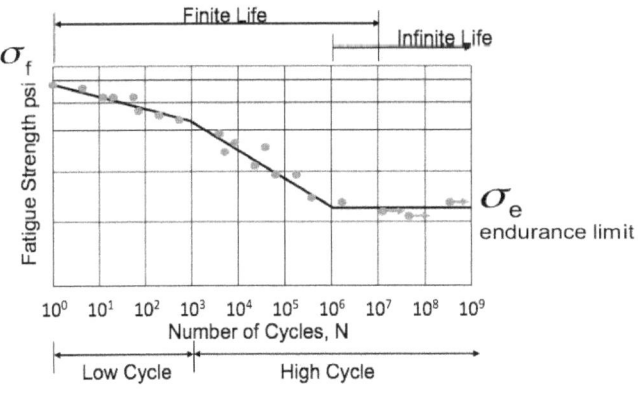

Figure 1-4 S-N Test Diagram

Design for Mechanical Fatigue

A typical S-N graph indicates that as the fatigue strength, measured by the fatigue test, decreases failure will occur at a longer but finite life of reversed loading cycles. At some level of reduced test specimen stress loading the specimen will no longer undergo fracture and demonstrates infinite life. This value of fatigue strength is defined as the endurance limit, σ_e, of the material. Most metallic materials behave in this fashion.

Fatigue Strength Equation

To provide a workable analytic representation for the S-N curve it is assumed that a log- log behavior between 10^3 and 10^6 cycles is acceptable. That is, σ_f, the fatigue strength is equal to a constant "a" times the number of cycles N raised to the "b" power (Figure 1-5). One condition used to define the constants is that at N equal to 10^3 cycles σ_f, the fatigue strength is equal to nine tenths of the ultimate static strength, σ_u. The second condition is that at N equal to 10^6 cycles σ_f will be equal to σ_e, the endurance limit. Beyond this number of cycles σ_f will always be σ_e.

That is $\sigma_f = aN^b$

where at $N = 10^3$ $\sigma_f = 0.9 \sigma_{ult}$

at $N = 10^6$ $\sigma_f = \sigma_e$ (endurance limit)

Solving for the constants "a" and "b" results in

$$\sigma_f = \frac{(0.9\sigma_u)^2}{\sigma_e} N^{-\frac{1}{3}\log\left(\frac{0.9\sigma_u}{\sigma_e}\right)}$$

Figure 1-5 Fatigue Strength Equation

Design for Mechanical Fatigue

The final expression for σ_f becomes 0.9 σ_u over σ_e times N raised to the minus 1/3 times the log of 0.9 σ_u divided by σ_e. This is somewhat complex but is quite workable and requires only two material properties, the static ultimate strength and the endurance limit.

Approximate Endurance Stress

When the endurance limit stress is not available from tests of specific materials the approximations in Figure 1-6 will result in a conservative design value for materials like wrought iron, steel and alloy steel.

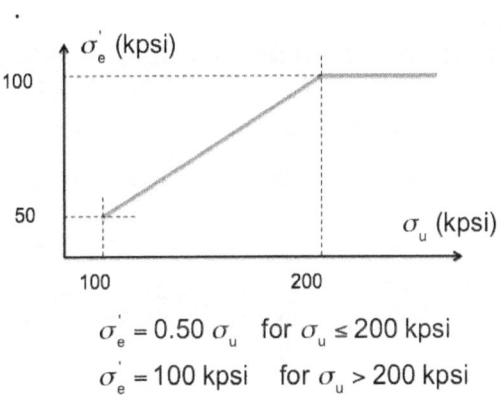

$\sigma_e' = 0.50\, \sigma_u$ for $\sigma_u \leq 200$ kpsi

$\sigma_e' = 100$ kpsi for $\sigma_u > 200$ kpsi

Figure 1-6 Approximate Endurance Stress

These values are designated σ_e' to distinguish them from actual measured endurance limit values.

Aluminum and copper alloys do not exhibit a defined endurance limit. However, these materials will eventually fail due to repeated loading. To come up with an "equivalent" endurance limit, designers typically use the value of the fatigue strength, σ_e', suggested in Figure 1-7 for aluminum and copper alloys at 10^8 cycles.

Aluminum Alloys (at 10^8 cycles)

$\sigma_e' = 0.4\, \sigma_{ut}$ for $\sigma_{ut} < 48$ Kpsi

Copper Alloys (at 10^8 cycles)

$\sigma_e' = 0.4\, \sigma_{ut}$ for $\sigma_{ut} < 35$ Kpsi

= 14 kpsi for all other vaues of σ_{ut}

Figure 1-7   Non Ferrous Endurance Limit

Endurance Limit Modifying Factors

In a real machine part the endurance limit stress will be reduced from that determined from a controlled fatigue test due to factors that differ from the idealized test condition. This is dealt with by the introduction of a number of correction factors K.

$\sigma_e' = K_a K_b K_c K_d K_e K_f\, \sigma_e$ = reduced endurance limit

K_a = surface factor (smoother → better)

K_b = size effect (smaller → better)

K_c = reliability

K_d = temperatur effect (varies with σ_u)

K_e = stress concentration effect

K_f = miscellaneous factors (residual stress, corrosion, fretting, etc.)

Figure 1-8   Endurance Modifying Factors

Design for Mechanical Fatigue

The effects generally included are surface condition, part size, reliability, operating temperature, stress concentration and other misc. effects. Each of these will be covered separately.

Surface Factor

The surface correction factor is given by the empirical equation K_a is equal to a constant "a" times σ_u, the ultimate strength, raised to the "b" power. Values of "a" and "b" are listed for a series of surface conditions for ferrous materials.

$$K_a = a\,\sigma_{ult}^{b}$$

Factors for ferrous materials, if surface is polished use $K_a = 1$.

Surface Finish	Factor "a" (kpsi)	Factor "b"
Ground	1.34	−0.085
Machined	2.70	−0.265
Cold Drawn	2.70	−0.265
Hot Rolled	14.4	−0.718
Forged	39.9	−0.995

Figure 1-9   Surface Factor Corrections

The ultimate strength, σ_{ult}, in this formula is in the units of kpsi. The smoother the surface the better. For nonferrous materials K_a is generally taken to be one.

Size Factor

The test specimens used to determine the fatigue strength and endurance limit are 0.30 inches in diameter. Tests performed with larger cross section diameters show a reduction in endurance limit of from 15 to 25 percent as indicated by diameter ranges listed in Figure 1- 10.

Design for Mechanical Fatigue

These endurance limit reductions are used for the size factor K_b reductions. They are also applicable for reversed axial loading.

For bending and torsion in circular shafts

$K_b = 1$ $d \leq 0.30$ in
$K_b = 0.85$ $0.30 < d \leq 2$ in
$K_b = 0.75$ $d > 2$ in.

Figure 1-10 Size Factor Correction

Reliability Factor

The information in Figure 1-11 for the reliability factor K_c is provided more as a guideline. It has not been sufficiently tested for recommended general use. The formula for K_c is given by one minus 0.08 times Z_R, a standardized variable that corresponds to the desired reliability R based on an eight percent standard deviation of the endurance limit. The table lists values of K_c for increasing vales of desired reliability.

$$K_c = 1 - 0.08\, z_R$$

Reliability R	Standardized Variable z_R	Reliability Factor K_c
50	0	1.0000
90	1.645	0.868
95	1.960	0.843
99	2.575	0.794
99.9	3.291	0.737
99.99	3.891	0.689

Figure 1-11 Reliability Factor

Design for Mechanical Fatigue

Temperature Factor

The temperature correction factor K_d is expressed as the ratio of the endurance stress at a specified operating temperature to the endurance stress at room temperature. Note that this correction factor is actually slightly greater than one until a temperature of 500 degrees Fahrenheit is reached. Then it begins to decrease quite rapidly.

$$K_d = \frac{\sigma_T}{\sigma_{RT}}$$

Temp., °F	σ_T/σ_{RT}	Temp., °F	σ_T/σ_{RT}
100	1.008	600	0.963
200	1.020	700	0.977
300	1.024	800	0.872
400	1.018	900	0.797
500	0.995	1000	0.690

Figure 1-12 Temperature Factor

Stress Concentration Factor

The following technique is used to determine how stress concentration effects due to geometry changes reduce the endurance limit. The factor K_e is defined as the reciprocal of a fatigue stress concentration factor K_f. The factor K_f is defined as the ratio of endurance limit of a notch-free specimen to that of the endurance limit obtained by testing a purposefully notched specimen. The sensitivity of this ratio gives rise to the definition of a notch sensitivity factor q defined as the ratio of $(K_f-1)/(K_t-1)$. Solving this equation for K_f gives the relation $1+ q(K_t-1)$ as illustrated in Figure 1-13.

$$K_e = \frac{1}{K_f}$$

Fatigue Stress Concentration Factor K_f

$$K_f = \frac{\text{endurance limit of notch-free specimens}}{\text{endurance limit of notched specimens}}$$

Notch Sensitivity q

$$q = \frac{K_f - 1}{K_t - 1} \quad K_t = \text{static geometric stress concentration factor}$$

Solving for K_f

$$K_f = 1 + q(K_t - 1)$$

Figure 1-13   Stress Concentration Factor

Typical values of the notch sensitivity q as a function of the notch radius on the test specimen and the ultimate static strength of a material are shown in Figure 1-14.

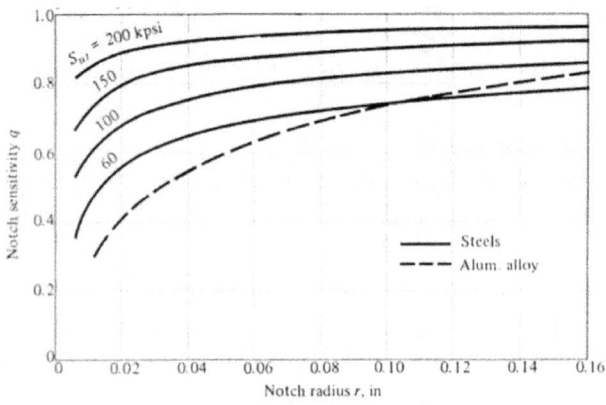

Figure 1-14 Sample Notch Sensitivity Chart

Design for Mechanical Fatigue

Miscellaneous Effect Factors

Other miscellaneous effects include residual stresses, corrosion, plating, cyclic frequency and fretting corrosion. The residual compressive surface stresses introduced by shot peening or cold working the surface actually increases the endurance limit. Corrosion conversely reduces the strength as roughness of the surface can lead to crack initiation. Plating or metal spaying has a similar effect and can reduce strength from 15 to 50 percent. The cyclic frequency of the rate of loading reversal is usually minimal and can be neglected. Fretting corrosion can be important to strength reduction at mechanical joints but is difficult to quantify.

Sample Application

A sample problem will be analyzed to demonstrate the quantitative application of these many correction factors.

A ¾ in. diameter solid shaft that operates at 1800 rpm under a constant bending load is machined from quenched and tempered steel with a yield strength of 75 kpsi and an ultimate strength of 97 kpsi. The shaft will operate at room temperature and contains a radial hole that introduces a geometric stress riser K_t of 1.5. If a reliability of 90 % is desired estimate the values of the modifying factors and determine the endurance limit for the part for the conditions specified. Assume a notch sensitivity based on a notch radius test specimen of $r = 0.04$ in.

Design for Mechanical Fatigue

Problem Solution

Since no endurance limit is given it is estimated from the ultimate strength of the material as one half of σ_u or 48.5 kpsi. The surface of the part has been machined so the surface factor K_a is equal to "a" times σ_u raised to the "b" power with "a" equal to 2.70 and "b" equal to minus 0.265 from Figure 1-9. It is calculated to be 0.802 in Figure 1-17.

Given: $\sigma_{ult} = 97$ kpsi, $\sigma_y = 75$ kpsi

Estimate test specimen endurance limit – σ_e

$$\sigma_e' = .50(97) = 48.5 \text{ kpsi}$$

Modifying factors:

Surface factor (machined surface)

$$K_a = a\, \sigma_{ult}^b \quad a = 2.70,\ b = -0.265$$

$$K_a = 270\,(97)^{-0.265} = 0.802$$

Figure 1-17 Surface Factor Correction

The size factor for a solid shaft in bending with the diameter d between 0.3 and 2 inches gives K_b as 0.85 from Figure 1 10.

Size factor (bending with .30 < d < 2):

$$K_b = 0.85$$

Reliability factor (90 %):

$$K_c = 0.868$$

Temperature Factor (room temperature)

$$K_d = 1$$

Figure 1-18 Size, Reliability & Temperature Factors

Design for Mechanical Fatigue

The desired reliability factor of 90 % results in K_c equal to 0.868 from Figure 1-11. The temperature factor K_d is one with room temperature operation.

Finally from the notch sensitivity graph in Figure 1-14 with r equal to 0.04 and σ_u = 97 kpsi the notch sensitivity q is approximately 0.75. This permits k_f to be calculated as 1.375. The stress concentration reduction factor K_e that is the reciprocal of K_f becomes 0.727

From the notch sensitivity graph with
$$r = 0.04 \text{ in.} \quad \sigma_{ut} = 97 \text{ kpsi}$$
$$\text{then} \quad q = 0.75$$
$$\text{then} \quad K_f = 1 + 0.75(1.5-1) = 1.375$$
$$\text{and} \quad K_e = \frac{1}{K_f} = \frac{1}{1.375} = 0.727$$

Figure 1-1   Stress Concentration Factor

All the correction factors are now multiplied together with the estimated endurance limit of 48.5 kpsi to give the reduced corrected endurance limit of 20.86 kpsi. It is observed that this is more than a fifty percent reduction. This seemingly drastic effect is a consequence of the correction factors all being multiplied together.

Mdified Endurnce Limit –
$$\sigma_e' = (K_a K_b K_c K_d K_e K_f)\sigma_e$$
$$\sigma_e' = \left[(0.802)(0.850)(0.868)(1)(0.727)\right](48.5)$$
$$\sigma_e' = 20.86 \text{ kpsi}$$

Figure 1-20   Final Reduced Endurance Stress

Design for Mechanical Fatigue

It should be kept in mind that the mathematical models proposed to describe the fatigue and endurance behavior of real materials are simplification of physical observation and statistical mechanical test results. They are at best approximations and should be treated as such. Their application provides a best estimate of a property involved in a very complex process of failure.

Design for Mechanical Fatigue

Chapter 2 – Fatigue Failure Prediction

Chapter 2 deals with the prediction of fatigue failure or finite design life as affected by the character and magnitude of the periodic loading on the machine part and its material properties.

Fluctuating Normal Stress States

Failure by fatigue is a consequence of the operation of a machine part subjected to a repetitive fluctuating stress level. One example of this form of loading is a completely reversed loading state where the applied normal stress oscillates from some compressive value to the same magnitude of tension and then repeats. Another example might be a repeated load that varies from zero to either some tension or compression and then repeats. The most general loading state is one where the stress varies from some compressive value to a tensile value of a different magnitude before repeating.

1. Complete Reversed Loading State
   $$-\sigma \rightarrow +\sigma \rightarrow -\sigma$$
2. Repeated Loading State
   $$0 \rightarrow \pm\sigma \rightarrow 0$$
3. General Loading State
   $$\sigma_{min} \rightarrow \sigma_{max} \rightarrow \sigma_{min}$$
   $$\sigma_{min} \leq \sigma_{mean} \leq \sigma_{max}$$

Figure 2-1   Fluctuating Stress State

Stress-Time Characterization

All of the example stress cycles in Figure 2-1 can be characterized by two generic parameters that have relevance in the limit to the materials properties that define static and dynamic failure. Recognizing that every cyclic loading creates some minimum stress and maximum stress this variation can be expressed in terms of a mean stress component and a stress amplitude component as shown in Figure 2-2. The mean stress component, σ_m, is simply the average of σ_{max} and σ_{min}. The stress amplitude, σ_a, is one half the difference between σ_{max} and σ_{min}. If σ_a is zero then the load is static and the important failure properties of the material are the yield and ultimate strength. If σ_m is zero then the loading is completely reversed and dynamic failure is defined by the endurance limit.

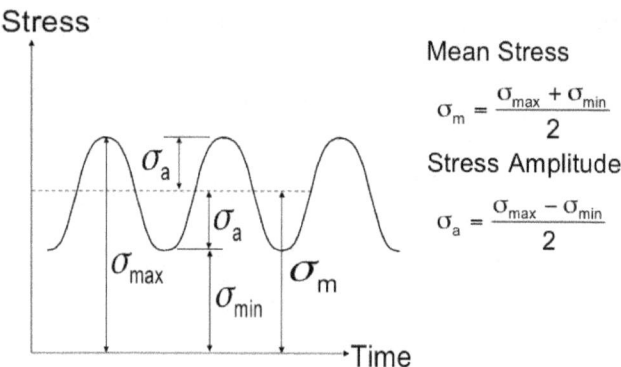

Figure 2-2 Stress Time Characterization

Design for Mechanical Fatigue

Mean/Fluctuating Stress Diagram

Illustrated in Figure 2-3 is the effect of stress amplitude versus mean stress on fatigue failure. Mean stress, σ_m, is plotted horizontally with the values of both the yield and ultimate strength indicated. The stress amplitude, σ_a, is plotted vertically with the endurance stress, σ_e indicated. If the stress amplitude is zero only a static mean stress would exist and failure would be predicted to occur at either the yield point or ultimate strength depending on the operational design requirement specified. Similarly, if the mean stress were zero only an alternating stress exists and failure in fatigue would be defined at the endurance strength of the material.

Failure test results of materials subjected to combinations of mean stress and stress amplitudes effectively all fall outside the line from σ_e to σ_u.

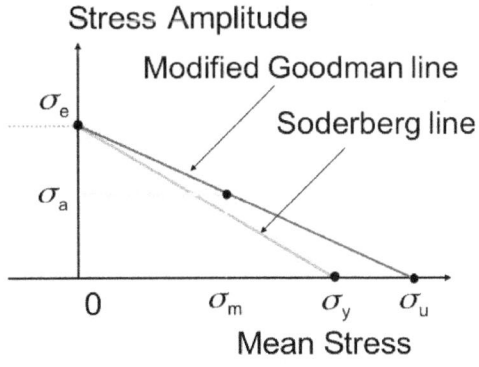

Figure 2-3 Mean/Fluctuating Stress Diagram

This is designated the Modified Goodman line and represents an appropriate criteria for a combination of σ_a and σ_m to predict fracture of the design. The line extending from stress amplitude σ_e to the mean stress σ_y is somewhat more conservative and is designated the Soderberg line. It is more representative of defining combinations of σ_a and σ_m when yielding is the criteria for failure under static load. The effect of negative values of σ_m will be dealt with later.

Fatigue Failure Theories

Listed in Figure 2-4 are the equations of the Goodman and Soderberg lines defining fatigue failure theory criteria for combinations of mean and alternating stress components. Included in each is a factor of safety, n. Given, the mean and alternating stresses to which a given design is exposed and the material from which it is produced, defining σ_e, σ_y and σ_u, these formula can be used to calculate the factor of safety or if failure is predicted.

$$\text{Goodman} - \quad \frac{\sigma_a}{\sigma_e} + \frac{\sigma_m}{\sigma_u} = \frac{1}{n}$$

$$\text{Soderberg} - \quad \frac{\sigma_a}{\sigma_e} + \frac{\sigma_m}{\sigma_y} = \frac{1}{n}$$

where n = factor of safety

Figure 2-4   Fatigue Failure Theories

Design for Mechanical Fatigue

Modified Fatigue Design Theory

To accommodate for the use of yielding as failure under static load as the more prevalent design specification the Modified Goodman theory is adjusted by the portion of the line from σ_a equal to σ_y to σ_m equal to σ_y. This is more conservative than the straight Goodman line to σ_u. Any combination of alternating stress, σ_a, and mean stress, σ_m, that falls on this modified line is considered to predict failure for positive values of σ_m. Test results for negative mean stress components indicate that it has little effect on when fatigue failure will occur under alternating stress. This is accounted for on the design theory graphic representation by the extended horizontal line from σ_e on the σ_a axis to the point of intersection with the line that connects σ_y on the σ_a axis to - σ_y on the σ_m axis on Figure 2-5.

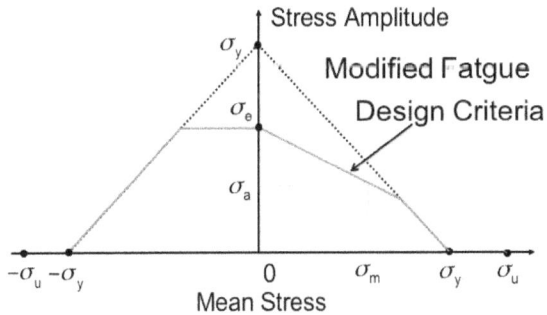

Figure 2-5 Modified Fatigue Design Theory

Design for Mechanical Fatigue

Sample Problem

Application of the modified fatigue design criteria is demonstrated with the following example problem. A 3.5 in. flat cantilever spring made from cold drawn steel strip is ground to cross sectional dimension's ½ in. by 1/16 in. It is loaded with a vertical end force that varies from 6 lbs. to 14 lbs. and operates at room temperature. A bevel at the fixed end introduces a stress concentration factor of $K_t = 1.20$. The tensile strength of the material is 85 kpsi. Calculate the factor of safety based on the following three conditions:

a. the mean stress remains constant
b. the stress amplitude remains constant
c. the ratio of the alternating stress to the mean stress remains constant

Determine the steel Endurance limit

$\sigma_e = 0.50\ \sigma_{ult}$

$\sigma_e = 0.50(85\ \text{kpsi}) = 42.5\ \text{kpsi}$

Determine the part Endurance limit

$\sigma_e' = K_a K_b K_c K_d K_e\ \sigma_e$

Surface factor (ground finish)

$K_a = a(\sigma_{ult})^b = 1.34(85)^{-0.085} = 0.918$

Size factor (rectangular cross section – 1/2" x 1/16")
since depth $h \leq 0.30"$
$K_b = 1$

Figure 2-6 Problem Calculation 1

Design for Mechanical Fatigue

In solving this problem the first task is to determine the endurance limit of the material using the correction factors covered in Chapter 1. This is illustrated in Figures 2-6 and 2-7.

> Reliability Factor
>   No reliability specified, therefor assume
> $K_c = 1$
> Temperature Factor
> $K_d = 1.008$
> Stress Concentration Factor (assume worst case)
> $K_e = 1/K_t = 1/1.2 = 0.833$
> Then
> $\sigma'_e = (.918)(1)(1)(1.008)(.833)42.5 = 32.54$ kpsi

Figure 2-7   Problem Calculation 2

Since no endurance limit is given it will be assumed an appropriate uncorrected value will be just half the ultimate strength. This gives σ_e equal to 42.5 kpsi. Next the correction factors K_a through K_e will be determined. With a ground finish the surface factor is calculated from Chapter 1 to be 0.918. With a rectangular cross section of ½ in. by 1/16 in. the height is less than 0.30 in. so the size factor K_b is equal to 1.

No specific reliability has been specified. There is no reason to reduce the endurance limit for this consideration so K_c is taken to be one. With the part operating at room temperature this dictates that the temperature factor K_d is equal to 1.008. The worst case

Design for Mechanical Fatigue

is assumed for the effect of the stress concentration at the base fillet since no information is provided on notch sensitivity. K_e becomes simply one over k_t or numerically 0.833. Multiplying the estimated endurance stress of 42.5 kpsi by all the correction factors gives a corrected endurance limit of 32.54 kpsi.

The mean and alternating loads and stresses are now calculated (Figure 2-8). The mean bending load is the average of the minimum and maximum loads, 6 plus 14 divided by 2 or 10 lbs. The mean stress is calculated from the bending stress formula M c/ I.

Loads and Streses

Mean Component

$$F_m = (6+14)/2 = 10 \text{ lb.}$$

$$\sigma = Mc/I, \quad I = bh^3/12 = \frac{(1/2)(1/16)^3}{12} = 1.02 \times 10^{-5} \text{ in.}^4$$

$$\sigma_m = \frac{(3.5 \times 10)(1/32)}{1.02 \times 10^{-5}} = 30.7 \text{ kpsi}$$

Alternating Component

$$F_a = (14-6)/2 = 4 \text{ lb.}$$

$$\sigma_a = \frac{4}{10} 30.7 = 12.3 \text{ kpsi}$$

Figure 2-8 Loads and Stresses

With the given dimensions and mean bending moment this becomes 30.7 kpsi. The alternating load component is the difference between the minimum and maximum loads divided by 2 or 4 lb. Taking the ratio of this load to the mean load multiplied by the mean bending stress gives an alternating bending stress of 12.3 kpsi.

In Figure 2-9 the large point on the modified Goodman diagram represents the calculated mean and alternating stresses. If the mean stress, σ_m, is held constant then the alternating stress, σ_a, can be increased to the value σ_a^a at point a before failure is predicted. The ratio of σ_a^a to σ_a is the factor of safety for the design for the first condition specified. Similarly if σ_a is held constant σ_m can be increased to σ_m^b at point b where failure is predicted. The ratio of σ_m^b over σ_m will be the factor of safety for this condition.

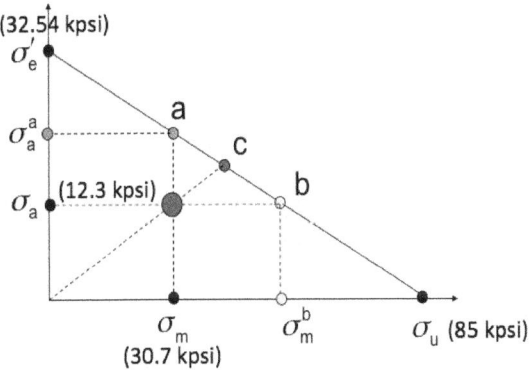

Figure 2-9   Goodman Diagram Solution

Finally if the ratio of σ_a to σ_m is held constant both stresses can be increased to point c where failure is predicted. Application of the Goodman failure theory

Design for Mechanical Fatigue

will give the factor of safety for the final condition of the problem.

By substituting σ_m at 30.7 kpsi into the Goodman failure theory along with the values of σ_e, 32.5 kpsi, and σ_u, 85 kpsi, σ_a^a can be determined. The value calculated is 20.8 kpsi. The factor of safety is given by the ratio of σ_a^a to σ_a as 1.69, Figure 2-10.

Part a:

σ_m remains constsnt, i.e. $\sigma_m = 30.7$ kpsi

From Goodman theory (at failure)

$$\frac{\sigma_a^a}{\sigma_e} + \frac{\sigma_m}{\sigma_{ult}} = 1 \Rightarrow \sigma_a^a = \sigma_e\left(1 - \frac{\sigma_m}{\sigma_{ult}}\right)$$

$$\sigma_a^a = (32.54)\left(1 - \frac{30.7}{85}\right) = 20.8 \text{ kpsi}$$

$$n = \frac{\sigma_a^a}{\sigma_a} = \frac{20.8}{12.3} = 1.69$$

Figure 2-10   Factor of Safety – Part a

The value of σ_a is now held constant and the procedure of the previous Figure is repeated using the Goodman failure theory. This gives a value for σ_m^b of 53 kpsi. The factor of safety for this condition is σ_m^b divided by σ_m giving a value of 1.73, Figure 2-11.

Design for Mechanical Fatigue

Part b:

σ_a remains constsnt, i.e. $\sigma_a = 12.3$ kpsi

From Goodman theory (at failure)

$$\frac{\sigma_a}{\sigma_e} + \frac{\sigma_m^b}{\sigma_{ult}} = 1 \Rightarrow \sigma_m^b = \sigma_{ult}\left(1 - \frac{\sigma_a}{\sigma_e}\right)$$

$$\sigma_m^b = (85)\left(1 - \frac{12.3}{32.7}\right) = 53.0 \text{ kpsi}$$

$$n = \frac{\sigma_m^b}{\sigma_m} = \frac{53.0}{30.7} = 1.73$$

Figure 2-11 Factor of Safety – Part b

For the solution to Part c the ratio of σ_a to σ_m is held constant at 0.401.

Part c:

σ_a / σ_m remains constsnt, i.e. $\sigma_a^c / \sigma_m^c = 12.3/30.7 = 0.401$

From Goodman theory (at failure)

$$\frac{\sigma_a^c}{\sigma_e} + \frac{\sigma_m^c}{\sigma_{ult}} = 1 \Rightarrow \sigma_m^c = \sigma_e \frac{1}{\left(\frac{\sigma_a^c}{\sigma_m^c} - \frac{\sigma_e}{\sigma_u}\right)}$$

$$\sigma_m^c = 32.54 \frac{1}{\left(\frac{12.3}{30.7} - \frac{32.7}{85}\right)} = 41.4 \text{ kpsi}$$

$$n = \frac{\sigma_m^c}{\sigma_m} = \frac{41.4}{30.7} = 1.35$$

Figure 2-12 Factor of Safety – Part c

This will also be the value of the ratio of σ_a^c to σ_m^c. Substituting this into the Goodman failure theory σ_m^c is calculated to be 41.4 kpsi. The factor of safety for this third condition is the ratio of σ_m^c to σ_m which gives a result of 1.35, Figure 2-12. This will be the minimum factor of safety for this design.

Check Modified Goodman

If the equations for the Goodman failure line and the static yield line of the design criteria in Figure 2-5 are set equal to each other the relationship between the value of σ_a and σ_m at the point of intersection of these two curves is obtained. This relation together with either the Goodman failure line or the static yield line can be used to calculate the values of σ_a and σ_m at the intersection point, Figure 2-13.

Goodman Line

$$\sigma_a / \sigma_e + \sigma_m / \sigma_u = 1$$

Yield Line

$$\sigma_a / \sigma_y + \sigma_m / \sigma_y = 1$$

Set equal and solve for $\sigma_a = f(\sigma_m)$

$$\sigma_a = \sigma_m \left（\frac{\sigma_e}{\sigma_u}\right)\left(\frac{\sigma_u - \sigma_y}{\sigma_y - \sigma_e}\right)$$

Figure 2-13 Failure Line Intersection

Before preceding a value of static yield stress is required. With none given in the original problem it will be assumed that σ_y is 70 kpsi to illustrate this checking procedure. At the intersection point σ_a is given by 0.153 σ_m. Substituting this into the equation for the static failure line results in a value of 60.7 kpsi for σ_m.

Consequently the value of σ_a at the intersection point is 9.3 kpsi, Figure 2-14.

Asume Yield Stress σ_y = 70 kpsi

at intersection point

$$\sigma_a = \sigma_m \left(\frac{32.5}{85}\right)\left(\frac{85-70}{70-32.5}\right) = 0.153\, \sigma_m$$

Substutute into Yield line

$\sigma_a/\sigma_y + \sigma_m/\sigma_y = 1 \Rightarrow \sigma_m = 70/0.153 = 60.7$ kpsi

and

$\sigma_a = 0.153\, \sigma_m = 9.3$ kpsi

Figure 2-15 Intersection Values

Illustrated in Figure 2-16, approximately to scale, is the modified fatigue design criteria diagram with all the pertinent problem stress values plotted. It is observed that the problem values for the mean and

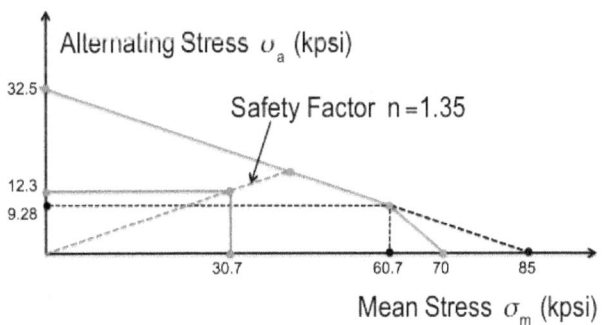

Figure 2-16 Modified Goodman Solution Graph

alternating stresses do in fact lie on the portion of the diagram where the application of the Goodman failure theory predicts the determination of the three factors of safety.

Fluctuating Torsional Fatigue

Up to this point the presentation has dealt only with the fatigue behavior of repetitive normal stress loading. Attention is now turned to fatigue due to application of a repetitive torsional stress loading. From the maximum static shear stress and distortion energy theories in Design for Static Mechanical Strength the yield stress in torsion, τ_y, is given as 0.50 and 0.577 of the static yield, σ_y, due to normal stress application.

From maximum –shear stress and distortion energy theories

$$\tau_y = 0.50\ \sigma_y$$

and

$$\tau_y = 0.577\ \sigma_y$$

From torsion fatigue testing

$$\tau_e = 0.50\ \sigma_e$$

and

$$\tau_e = 0.577\ \sigma_e$$

Figure 2-17 Torsional Yield & Endurance Stress

The results of the tests of torsional loaded specimens to determine the endurance limit in shear indicate that a similar relationship exists between

Design for Mechanical Fatigue

endurance limits. A reasonable assumption is the shear endurance limit can be taken to be 0.50 or 0.577 of the normal stress endurance limit σ_e. The more conservative approach would be to use the 0.50 value.

Torsion Fatigue Design Criteria

Fatigue test results with combinations of applied mean and alternating components of torsional loading indicate that the value of the mean stress has little effect on when fatigue failure takes place. This results in the horizontal line on the torsion fatigue design criteria diagram.

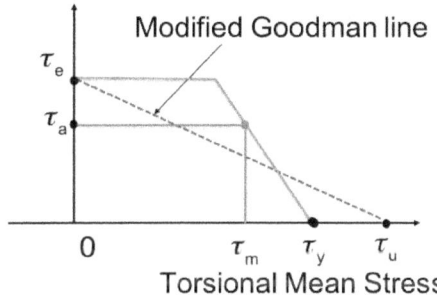

Figure 2-18   Torsional Fatigue Diagram

For higher values of mean torsional stress the static failure curve is used to define the boundary of the failure region. For comparison the modified Goodman line is included in Figure 2-18.

Torsional Fatigue Formula

Design for Mechanical Fatigue

Two mathematical formulas define the torsional fatigue behavior shown in Figure 2-19. For true fatigue behavior the ratio of the endurance shear limit to the alternating shear stress loading is equal to one over the factor of safety. For the static failure line the ratio of the alternating shear, τ_a, to the yield shear, τ_y plus the ratio of the mean shear, τ_m, to the yield shear, τ_y is equal to one over the factor safety.

For Fatigue Failure
$$\frac{\tau_a}{\tau_e} = \frac{1}{n}$$

For Static Failure
$$\frac{\tau_a}{\tau_y} + \frac{\tau_m}{\tau_y} = \frac{\tau_{max}}{\tau_y} = \frac{1}{n}$$

Point of Intersection
$$\tau_m = \tau_a \left(\frac{\tau_y}{\tau_e} - 1 \right)$$

Figure 2-19 Torsional Factors of Safety

It is observed that the alternating shear plus the mean shear are also equal to the maximum shear τ_{max} so that τ_{max} over τ_y is also equal to one over the factor of safety. The point of intersection of the fatigue and static lines is given by τ_m equal to τ_a times the quantity τ_y over τ_e minus one.

Chapter 3 – Multiple Stress Loading

Chapter 3 of Design for Mechanical Fatigue deals with predicting the fatigue behavior when the loading consists of a multiple stress state that can include both normal and shear mean and alternating stress components. An example application of the multiple step solution process is included.

Combined Loading Analysis Process

Figure 3-1 contains the first three steps of a six-step process for analyzing the fatigue behavior resulting from a loading that consists of both mean and alternating normal and shear stresses. The first step is to determine the mean and alternating stress components for all applied loads and combine them into σ_x, sσ_y and τ_{xy} mean and alternating stress sets.

1. Determine mean and alternating stress components for all applied loads and combine as appropriate, i.e.
$(\sigma_x^m, \sigma_y^m, \tau_{xy}^m), (\sigma_x^a, \sigma_y^a, \tau_{xy}^a)$

2. Apply fatigue stress concentration factors to alternating components of all resulting stresses.

3. Calculate principal stresses (σ_1^m, σ_2^m) from mean stress components and principal alternating stress (σ_1^a, σ_2^a) components using Mohr's circle

Figure 3-1 Combined Loading Analysis – Steps 1-3

The second step is to apply stress concentration factors to the alternating components of all resulting stresses. This will result in a more conservative final failure criteria evaluation. The third step is to calculate the principal stresses σ_1^m and σ_2^m for the mean stress components. In a similar fashion the principle alternating stresses σ_1^a and σ_2^a are determined for the alternating stress components.

The fourth step (Figure 3-2) is to calculate an effective Von Misses stress σ_a^e for the alternating principle stresses using the generic formula presented. An effective Von Misses mean stress σ_m^e is also calculated for the principle mean stress components. In step five the material fatigue property is established by using the endurance limit in bending that is corrected for all reducing factors except stress concentrations.

4. Calculate effective alternating stress (σ_a^e) and effective mean stress (σ_m^e) using

$$(\sigma_x^e) = \sqrt{\left(\sigma_1^x\right)^2 - \left(\sigma_1^x \sigma_2^x\right) + \left(\sigma_2^x\right)^2} \quad \text{(Von Mises stress)}$$

5. For part material properties use endurance limit in bending corrected for all effects except stress concentration and tensile strength for ultimate strength.

6. Apply Goodman fatigue analysis using effective (σ_a^e) alternating stress (σ_m^e) effective mean stress.

Figure 3-2 Combined Loading Analysis Steps 4-6

Also the tensile strength of the material is used for the ultimate strength. Finally in step six the Goodman fatigue analysis is applied using the effective Von

Misses stress σ_a^e for the alternating stress and the effective Von Misses stress σ_m^e for the mean stress component. Care must be exercised in combining stresses created by the loading system into proper mean and fluctuating stress components prior to application of this procedure.

Generic Loaded Rotating Shaft

To demonstrate the determination of correct alternating and mean stress components for a complex loading system the generic problem of a rotating shaft subjected to constant bending, torsion, direct shear and tension will be analyzed. To do this properly the stress states generated on an element of the shaft as it passes through positions A, B, C and D must be considered (Figure 3-3).

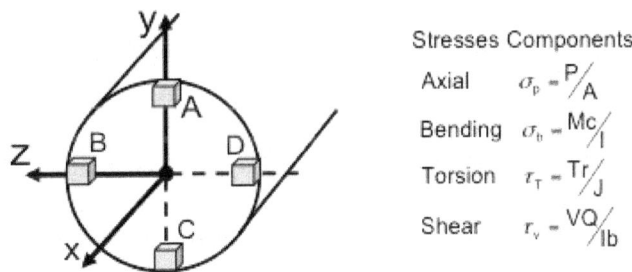

Figure 3-3 Generic Rotating Shaft

The axial stress component σ_p will be given by the axial load P divided by the cross sectional area A. The bending stress σ_b will be calculated from simple beam theory with M the bending moment multiplied by the dimension c divided by I the cross sectional moment of inertia. The bending moment is assumed to act about the z-axis to produce tension at the top of the shaft.

Torsion about the x-axis will produce a shear stress τ_r given by the applied torque T multiplied by radial position r divided by J the polar moment of inertia. Finally, the direct shear τ_v is given by the shear force V multiplied by the first moment Q dived by the product of the cross sectional inertia I times the width of the section b. The direct shear V is assumed to act in the negative y direction.

Element Stresses at A & B

At the top of the cross section at position A (Figure 3-4) there are normal stress components σ_b and σ_p acting on the element in the x direction due to both the bending and the axial load which are additive. Both σ_z and σ_y are zero. There is no direct shear stress τ_v at this location since Q will be zero but there is a shear stress due to the torsion τ_r.

At location B (Figure 3-4) on the mid-plane of the cross section the bending stress σ_b will be zero but the axial stress σ_p will still act in the x direction. A direct shear τ_v will exist as will a torsional shear stress τ_r. They will be additive at this location assuming the torque acts clockwise about the x direction.

Design for Mechanical Fatigue

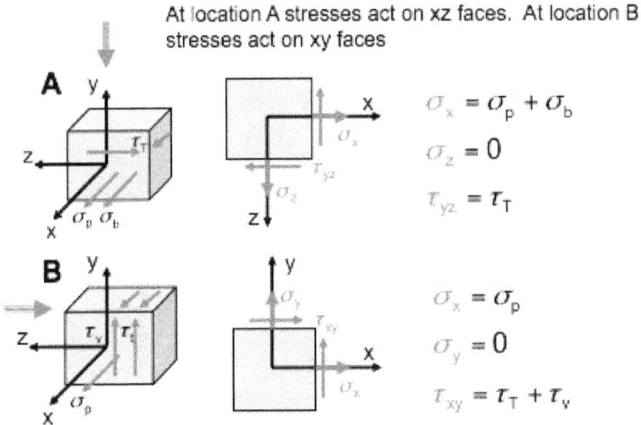

Figure 3-4 Element Stresses at A and B

Element Stresses at C & D

At the bottom of the shaft at location C (Figure 3-5) the bending stress σ_b will be compressive so the net normal stress will be the axial stress σ_p minus σ_b. There will only be shear stress due to torsion τ_t since again the direct shear is zero.

At location D (Figure 3-5) the bending stress σ_b is again zero leaving only the axial stress σ_p as the normal stress component. There will now be a direct shear stress τ_v at this location but it will be in the opposite direction to the shear stress due to torsion τ_t. (Figure 3-5)

Design for Mechanical Fatigue

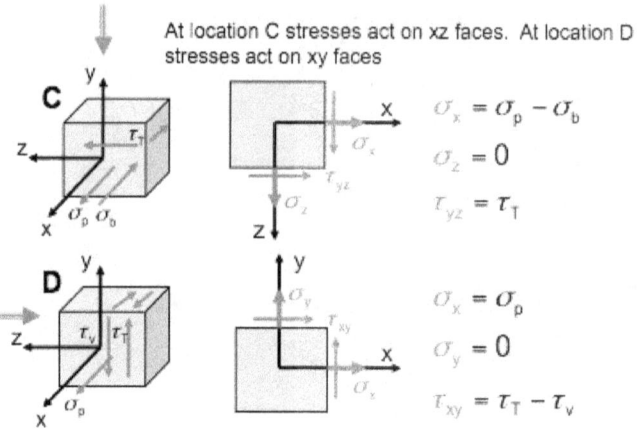

Figure 3-5   Element Stresses at C and D

Mean and Alternating Stresses

The mean normal stress will be the average of the max and min normal stresses acting on the element as it rotates around the x-axis. Substituting these values from the stresses at the four location results in σ_m simply equal to σ_p generated by the constant axial load. The alternating normal stress component is the difference between the maximum and minimum normal stress values dived by two. Carrying out this calculation results in σ_a equal to the maximum bending stress σ_b (Figure 3-6).

In a similar fashion the mean shear stress component is calculated to be simply the value of max shear stress τ_t generated by the delivered torque. The alternating shear stress component becomes the maximum value of the direct shear τ_v acting at point B and D. (figure 3-6)

Normal Stresses

$$\sigma_m = \frac{\sigma_{max} + \sigma_{min}}{2} = \frac{(\sigma_b + \sigma_p) + (\sigma_p - \sigma_b)}{2} = \sigma_p$$

$$\sigma_a = \frac{\sigma_{max} - \sigma_{min}}{2} = \frac{(\sigma_b + \sigma_p) - (\sigma_p - \sigma_b)}{2} = \sigma_b$$

Shear Stress

$$\tau_m = \frac{\tau_{max} + \tau_{min}}{2} = \frac{(\tau_T + \tau_v) + (\tau_T - \tau_v)}{2} = \tau_T$$

$$\tau_a = \frac{\tau_{max} + \tau_{min}}{2} = \frac{(\tau_T + \tau_v) - (\tau_T - \tau_v)}{2} = \tau_v$$

Figure 3-6 Mean and Alternating Stresses

Graphic Representation

The variation of both the normal stresses and shear stresses acting on the element as it rotates about the x-axis to the four position A, B, C and D are illustrated in Figure 3-7.

In terms of the maximum and minimum values of these stress variations the mean and alternating normal and shear stress components are clearly observed to be the values calculated above.

Design for Mechanical Fatigue

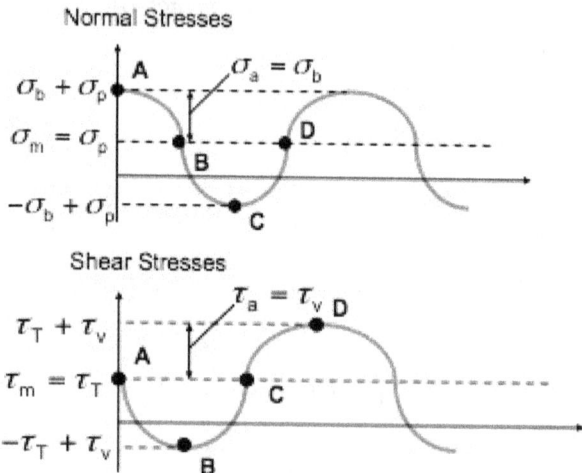

Figure 3-7 *Graphic Alternating Stresses*

Sample Problem

The results of the previous loading analysis together with the process of fatigue analysis for combined loading will be demonstrated with the following problem. A rotating shaft transmits a constant torque that creates a maximum shear stress of 8 kpsi. The shaft is also subjected to a constant axial force that produces a tension of 10 kpsi. In addition the shaft carries a bending load that results in a maximum alternating bending stress of +/- 23 kpsi. Apply a geometric stress concentration factor of $K_t=1.2$ Determine the factor of safety for this shaft if its material has a tensile strength of 75 kpsi and its corrected endurance limit is 47 kpsi.

Design for Mechanical Fatigue

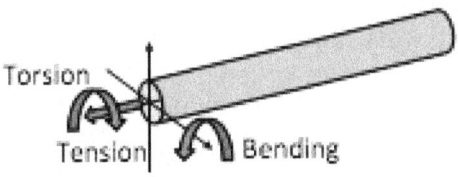

Figure 3-8   Rotating Shaft Loading

Mean/Alternating Stress Values

The two-dimensional stress state that will act on an element of the shaft as it rotates is illustrated by the graphic at the top left of Figure 3-9.   Step one of the fatigue analysis process is to establish the values of the combined mean and alternating normal and shear stress components, σ_a, σ_m, τ_a and τ_m.

Step 1

$\tau = 8$ kpsi
$\sigma_b = \pm 23$ kpsi
$\sigma_t = 10$ kpsi

$$\sigma_a = \frac{33-(-13)}{2} = 23 \text{ kpsi}$$

$$\sigma_m = \frac{33+(-13)}{2} = 10 \text{ kpsi}$$

$\tau_m = 8$ kpsi, $\tau_a = 0$ kpsi

Step 2   Correct Alternating Stress Components for Geometric Stress Concentration

$\sigma_a = 23 \times K_t = 23 \times 1.2 = 27.6$ kpsi

Step 3   Correct Alternating Axial Stress

no correction required

Figure 3-9   Mean and Alternating Stress Values

From the given stress state σ_a will simply be σ_b which is 23 kpsi. σ_m will be the value of the constant axial stress of 10 kpsi.   There will be no τ_a since there is no direct shear and σ_m will be the constant torsion shear stress of 8 kpsi. Step 2 corrects the alternating stress component by the given stress concentration value of 1.2 to give a final σ_a of 27.6 kpsi.

Principal Mean and Alternating Stress Values

Step 3 in the multiple stress fatigue analysis process is calculating the principal normal mean and alternating stresses. There are both mean normal and shear stresses from the applied stress state that contribute to the principal normal mean stresses.

Principal Components

$$\sigma_{1,2}^m = \sigma_m/2 \pm \sqrt{(\sigma_m/2)^2 + \tau_m^2} = 5 \pm \sqrt{25 + 64}$$

$$\sigma_{1,2}^m = 5 \pm 9.43$$

$$\sigma_1^m = 14.43 \text{ kpsi}, \quad \sigma_2^m = -4.43 \text{ kpsi}$$

$$\sigma_{1,2}^a = \sigma_a/2 \pm \sqrt{(\sigma_a/2)^2 + \tau_a^2} = 13.8 \pm \sqrt{13.8^2 + 0}$$

$$\sigma_{1,2}^a = 13.8 \pm 13.8$$

$$\sigma_1^a = 27.6 \text{ kpsi}, \quad \sigma_2^a = 0 \text{ kpsi}$$

Figure 3-10   Principal Mean and Alternating Values

The two values calculated for σ_1^m and σ_2^m in Figure 3-10 are 14.43 kpsi and -4.43 kpsi respectively. In calculating the principal normal alternating stress values it is recognized that there is no contribution from the applied stress state of an alternating shear component. Therefore sigma σ_1^a, σ_2^a is simply +/- 13.8 kpsi. This results in σ_1^a equal to 27.6 kpsi while σ_2^a is zero.

Effective Von Misses Stresses

The effective von misses stresses are now calculated using the principal mean and alternating stress values determined in Figure 3-10. This calculation is made using the generic formula in Figure 3-2 to give values of 17.07 kpsi for σ_m^e and 27.6 kpsi for σ_a^e since σ_2^a is zero.

Effective Stresses

$$\sigma_e^m = \sqrt{(\sigma_1^m)^2 - (\sigma_1^m)(\sigma_2^m) + (\sigma_1^m)^2}$$

$$\sigma_e^m = \sqrt{(14.43)^2 - (14.43)(-4.43) + (-4.43)^2} = \sqrt{291}$$

$$\sigma_e^m = 17.07 \text{ kpsi}$$

$$\sigma_e^a = \sqrt{(\sigma_1^a)^2 - (\sigma_1^a)(\sigma_2^a) + (\sigma_1^a)^2}$$

$$\sigma_e^a = \sqrt{(27.6)^2 - (27.6)(0) + (0)^2}$$

$$\sigma_e^a = 27.6 \text{ kpsi}$$

Figure 3-11   Effective Von Misses Stresses

Goodman Diagram

A Goodman diagram is now constructed from the given corrected endurance stress of 47 kpsi on the vertical effective alternating stress axis to 75 kpsi for σ_u on the effective mean stress axis in Figure 3-12. The

Design for Mechanical Fatigue

calculated values of σ_m^e of 17.1 kpsi and σ_a^e of 27.6 kpsi are plotted on the two axes. A diagonal is passed through the point where these two values intersect. Assuming that failure will not occur with this stress state the diagonal is extended until it reaches the dot on the Goodman line predicting failure. By determining the value of $\sigma_a^{e'}$ to the left of this point and taking its ratio with σ_a^e of 27.6 kpsi the factor of safety for this loading condition can be determined.

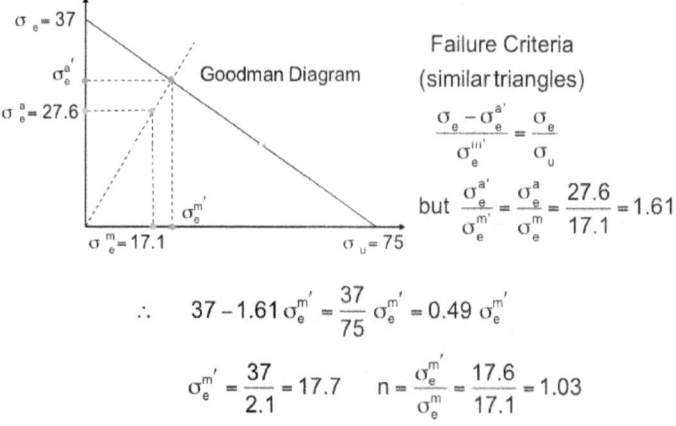

Figure 3-12 Goodman Diagram

From the geometry of similar triangles the value of $(\sigma_e - \sigma_e^{a'})/\sigma_e^{m'}$ will be equal to the ratio of σe to σ_u. Also recognizing that $\sigma_e^{a'}/\sigma_e^{m'}$ is equal to σ_e^a/σ_e^m is calculated as 27.6 divided by 17.1 or 1.61. Then the fatigue failure criteria equation can be solved for $\sigma_e^{m'}$ as 17.6 kpsi. The factor of safety, $\sigma_e^{m'}/\sigma_e^m$ is then calculated to be 1.03. This low factor of safety indicates that the shaft is very close to failure by fatigue with the loading as specified.

Design for Mechanical Fatigue

Design for Mechanical Fatigue

Chapter 4 – Partial Fatigue Damage

Chapter 4 of Design for Mechanical Fatigue deals with cumulative fatigue damage. It covers the Palmgren-Miner method of modeling cumulative damage and the Mason modification of this method. An application of both is included to calculate total part life and reduction of the effective endurance limit under conditions of partial fatigue damage.

Fatigue Damage Accumulation

The simplest damage accumulation model predicts that failure will occur when the linear sum of a series of minor individual damage occurrences total some critical value. Consider that d_i represents the damage created by some occurrence and D is the total damage required to produce failure. The sum of d_i over N occurrences equal to D defines mathematically when failure will occur.

Linear Model

d_i = damage created by some occurence

D = total damage required to produce failure

then failure occurs when

$$\sum_{i=1}^{N} d_i = D \quad \text{or} \quad \sum_{i=1}^{N} \frac{d_i}{D} = 1$$

where $\frac{d_i}{D}$ = fractional damage of the ith contribution

Figure 4-1 Liner Damage Model

This equation can be rewritten in the form the summation of the ratios of d_i over D for N occurrences is equal to one, see Figure 4-1. In this formulation the ratio of d_i to D represents the percentage or fractional damage of that contribution.

Palmgren-Miner Summation Theory

The Palmgren–Miner Summation Theory for cumulative fatigue damage is based on this simple linearly additive model of occurrence effects. The theory proposes that the percentage or fractional damage in fatigue from reversed loading at a specified alternating stress level is the ratio of the number of load cycles applied divided by the total number of cycles available to failure at that alternating stress level.

Palmgren–Miner Summation Theory
(for reversed loading)

$$\frac{n_1}{N_1} + \frac{n_2}{N_2} + \frac{n_3}{N_3} + \cdots = C \quad 0.7 \leq C \leq 2.2$$

n_i = cycles of reversed stress σ_i

N_i = total life cycles at reversed stress σ_i

Usual practice (conservative) –

$C = 1$

Figure 4-2   Cumulative Fatigue Damage

The sum of these percentages is set equal to some constant C. Experimental test results indicate that the constant C lies between 0.7 and 2.2. For simplicity the constant is generally taken to be one.

Design for Mechanical Fatigue

Sample Problem

To illustrate he application of the Palmgren–Miner theory the following problem will be solved in detail. A part with a tensile strength of 100 kpsi and an endurance limit of 45 kpsi is subjected to a reversed normal stress of 69 kpsi for 3500 cycles. For this set of conditions find the following:

1. The remaining life of the part if the stress remains at 69 kpsi.
2. The remaining life if the stress is reduced to 45 kpsi from 69 kpsi, and
3. The reduced endurance limit after being subjected to 69 kpsi for 3500 cycles.

Log- Log SN Diagram

It is assumed that the generic log–log SN diagram, shown in Figure 4-3, governs the fatigue From Chapter 1 of Design for Mechanical Fatigue. This is represented by a straight line from 0.9 σ_u on the vertical alternating stress axis at 1000 cycles to the endurance limit σ_e on the horizontal axis at 10^6 cycles. Beyond that point the curve remains horizontal. At a particular reversed or alternating stress level like σ_1 the life of the material at which fatigue failure occurs is indicated by N_1 on the cycle scale.

To determine the solution to part 1 of the problem the value of N_1 needs to be determined following σ_1 being applied for n_1 cycles.

Design for Mechanical Fatigue

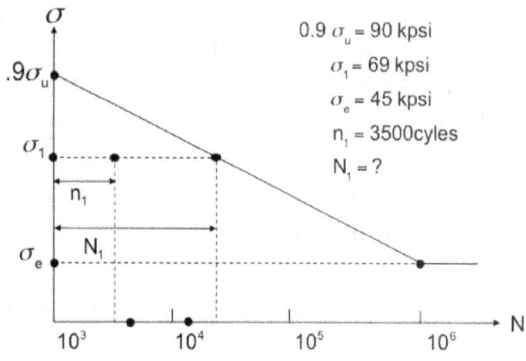

Figure 4-3 Sample Problem SN Diagram

Part 1 Solution

The easiest way to determine cap N 1 is to make use of the two similar triangles shown on the SN diagram in Figure 4-4. On the triangle in the lower right both the height and the base are known. On the triangle in the upper left the height is known but the base is the unknown value of N1.

By setting the ratio of the height to the base equal for these two similar triangles an equation is obtained permitting N1 to be determined numerically. From the similar triangles in Figure 4-4 the ratio of the heights to the bases of the two triangles are set equal to one another recognizing that the SN curve is represented on a log-log plot.

Design for Mechanical Fatigue

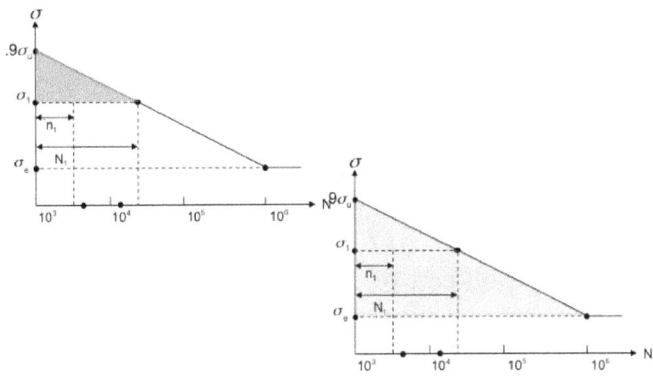

Figure 4-4 Similar Triangles with N_1 Unknown

From this relationship N1 is determined to be $10^{4.11}$ or 12,590 cycles.

From Similar Traingles on SN Diagram

$$\frac{\log(0.9\,\sigma_u) - \log(\sigma_1)}{\log(N_1) - \log(10^3)} = \frac{\log(0.9\,\sigma_u) - \log(\sigma_e)}{\log(10^6) - \log(10^3)}$$

$$\frac{4.95 - 4.84}{\log N_1 - 3} = \frac{4.95 - 4.65}{6 - 3}$$

$$\log N_1 = 3 + 3\left(\frac{0.11}{0.3}\right) = 4.11$$

$$N_1 = 10^{4.11} = 12{,}590 \text{ cycles}$$

Remaining life = 12,590 – 3500 = 9,090 cycles

Figure 4-5 Solution for Cycle Life N_1

The remaining life of the part at σ_1 = 69 kpsi is then the difference between this total life N_1 and the initial applied cycles n_1 of 3500 cycles. This gives a value of 9,090 cycles that still remains at the applied stress of 69 kpsi.

Part 2 Solution

The Palmgren Miner theory is now applied in Figure 4-6 to determine the remaining life of the part if after being subject to σ_1 of 69 kpsi for 3500 cycles the stress level is reduced to 45 kpsi, the endurance limit. This condition is represented by the equation that n_1/N_1 plus n_2/N_2 is equal to one. In this relationship the only unknown is little n_2 since it is known from the SN diagram that N_2 is 10^6 cycles. Solving for little n_2 gives a value of 722,000 cycles. This is some 280,000 cycles less than the part would have sustained had it been subjected to σ_e for its entire life.

From Palmgren–Miner Theory

$$\frac{n_1}{N_1} + \frac{n_2}{N_2} = 1$$

$$n_2 = \left(1 - \frac{n_1}{N_1}\right) N_2 = \left(1 - \frac{3{,}500}{12{,}590}\right) 10^6$$

$$n_2 = .772 \times 10^6 = 722{,}000 \text{ cycles}$$

Figure 4-6   Solution for Remaining Cycle Life

Part 3 Solution – Damaged SN Diagram

To determine the solution to the third part of the problem it is first necessary to establish what is

Design for Mechanical Fatigue

referred to as the damaged SN diagram (Figure 4-7). Palmgren-Miner proposes this is created by first plotting on the σ_1 bevel, a dotted horizontal line, a point designated "N_1'" defined by the difference between N_1 and n_1 on the horizontal cycle axis. This is indicated by a light gray dot on the diagram. A straight line is then constructed parallel to the original SN diagram through this point. Where this new dotted SN line intersects a vertical through 10^6 cycles defines the level of the reduced endurance limit σ_e^1 of the partially damaged part.

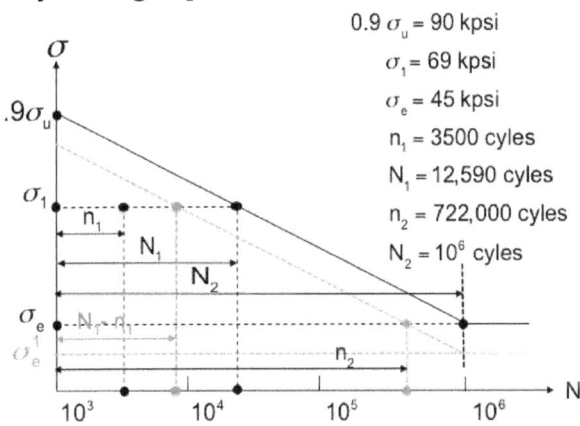

Figure 4-7   Palmgren-Miner   Damaged SN Diagram

Similar triangles are again used to determine the value of the reduced endurance limit σ_e^1 on the damaged SN diagram (Figure 4-8).

Design for Mechanical Fatigue

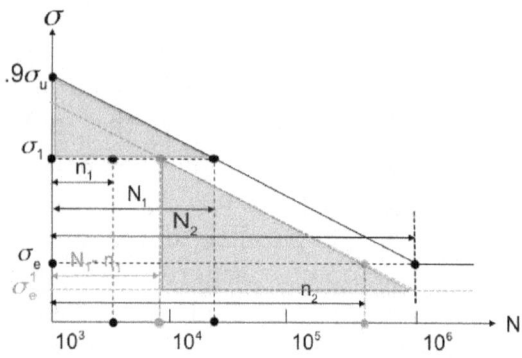

Figure 4-8 Similar Triangles on P-M SN Diagram

Both the base and the height of the upper left triangle are known. On the lower right triangle the base is known but the height is dependent on the value of the reduced endurance limit σ_e^1. Setting these base to height ratios equal on the two similar triangles permits the reduced endurance limit to be determined.

The equation relating the ratio of the height to the base of the similar triangles from Figure 4-8 recognizing that the SN diagram is a log–log plot appears in Figure 4-9. The only unknown in this equation is the reduced endurance limit, σ_e^1. Solving for this unknown results in the log of σ_e^1 equal to 4.636. The final numerical value of the reduced endurance limit is 43,250psi. Any stress level applied to the part below this value following the initial damage created by 69 kpsi for 3500 cycles would result in an infinite life for the part.

Design for Mechanical Fatigue

From Similar Traingles on Damaged SN Diagram

$$\frac{\log(0.9\,\sigma_u) - \log(\sigma_1)}{\log(12{,}590) - \log(10^3)} = \frac{\log(\sigma_1) - \log(\sigma_e^1)}{\log(10^6) - \log(9{,}090)}$$

$$\frac{4.95 - 4.84}{4.10 - 3} = \frac{4.95 - \log(\sigma_e^1)}{6 - 3.96}$$

$$\log \sigma_e^1 = 4.84 - 2.04\left(\frac{0.11}{1.10}\right) = 4.636$$

$$\sigma_e^1 = 10^{4.636} = 43{,}250 \text{ cycles}$$

Figure 4-9   Reduced Endurance Limit

Mason Modification

A modification of the Palmgren Miner theory proposed by Mason is that at 10^3 cycles it does not appear reasonable to reduce the ultimate strength of the material on the damaged SN diagram. This reorients the damaged SN diagram to now pass through the point $0.9\,\sigma_u$ and the previously established dot at coordinates σ_1 at $N_1 - n_1$.

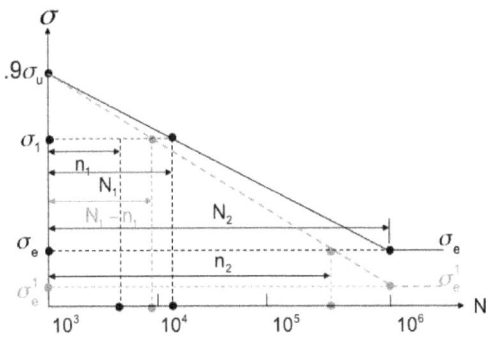

Figure 4-10   Mason Modified SN Diagram

The result of this change is that the remaining life at σ_e designated by n_2 and the reduced endurance limit σe^1 at 10^6 cycles will be lower than those obtained applying the Palmgren- Miner method (Figure 4-10).

Part 1 Solution Modified

The two similar triangles shown on the Mason modified SN diagram in Figure 4-11 can again be used to determine the value of n2 which will be the reduced life at σ_e following the damage produced by the 69 kpsi stress for 3500 cycles.

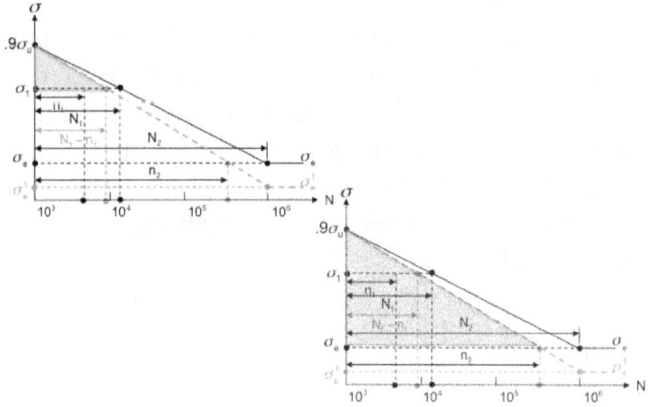

Figure 4-11 Similar Triangles on Modified SN

Using the same procedure of equating the height to the base of similar triangles for determining n_2 by the Palmgren-Miner Method the remaining life with the Mason modification is reduced from the previous 722,000 cycles to 407,400 cycles (Figure 4-12). This represents a reduction of some six hundred

Design for Mechanical Fatigue

thousand cycles from that of the original endurance limit, σ_e, life of the material.

From Similar Traingles on Modified SN Diagram

$$\frac{\log(0.9\,\sigma_u) - \log(\sigma_1)}{\log(N_1 - n_1) - \log(10^3)} = \frac{\log(0.9\,\sigma_u) - \log(\sigma_e)}{\log n_2 - \log(10^3)}$$

$$\frac{4.95 - 4.84}{3.96 - 3} = \frac{4.95 - 4.65}{\log n_2 - 3}$$

$$\log n_2 = 3 + 0.96\left(\frac{0.30}{0.11}\right) = 5.61$$

$$n_2 = 10^{5.61} = 407{,}380 \text{ cycles}$$

Figure 4-12   Reduced Cycle Life at σ_e

Part 2 Solution Modified

Next the two similar triangles shown in Figure 4-13 on the damaged Mason modified SN diagram will be used to determine how much further the damaged endurance limit is reduced with the Mason modification.

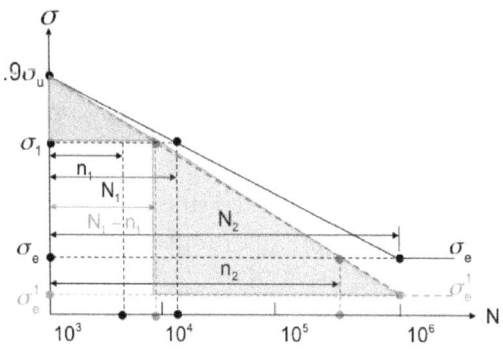

Figure 4-13   Triangles for Calculating $\sigma e1$

Part 3 Solution Modified

By setting the ratio of the heights to the bases of these triangles equal to one another the only unknown in these dimensions is the new reduced endurance limit σ_e^1. The mathematical calculation of the reduced endurance σ_e^1 is detailed in Figure 4-14. The calculations proceed the same as in Figure 4-9 for the calculation of the reduced endurance limit using the Palmgren-Miner method. With the Mason modification of the damaged SN diagram the reduced endurance limit is calculated to be 40,700 psi.

From Similar Traingles on Damaged SN Diagram

$$\frac{\log(0.9\,\sigma_u) - \log(\sigma_1)}{\log(N_i - n_1) - \log(10^3)} = \frac{\log(0.9\,\sigma_u) - \log(\sigma_e^1)}{\log(10^6) - \log(10^3)}$$

$$\frac{4.95 - 4.84}{3.96 - 3} = \frac{4.95 - \log(\sigma_e^1)}{6 - 3}$$

$$\log \sigma_e^1 = 4.95 - 3\left(\frac{0.11}{0.96}\right) = 4.61$$

$$\sigma_e^1 = 10^{4.61} = 40,738 \text{ cycles}$$

Figure 4-14 Calculation of Mason σe^1

Summary of Results

Tabulated in Figure 4-15 are the results of the solution of the sample problem using both the Palmgren-Miner method and Mason modification to determine limited life following initial damage and the reduced endurance limit. It is observe that both methods predict greater percentage reduction in life as compared to the endurance limit.

Design for Mechanical Fatigue

Original Material: Ultimate Stress – 100,000 psi
Endurance Limit – 45,000 psi

	Stress Level 1	n_1 cycles	N_1 cycles	Stress Level 1	n_2 cycles	Endurance limit
Palmgrem Miner	69,000	3,500	12,590	45,000	772,000	42,250
Msaon	68,000	3,500	12,590	45,000	407,380	40,738

Figure 4-15 Summary of Solution Results

As a word of caution there can be many more effects that impact partial damage than considered in these simplified models. Care should be used in using these methods for detailed design considerations. They do however provide guidance on the magnitude of how accumulated fatigue damage can affect a specific application.

Design for Mechanical Fatigue

www.ingramcontent.com/pod-product-compliance
Lightning Source LLC
Chambersburg PA
CBHW061206180526
45170CB00002B/985